LDS Family History
The EASY Way

The Savior *did* say **it would be easy**

(Matthew 11:30)

by

Chuck Call

ISBN: 1-4107-7727-8 (e-book)
ISBN: 1-4107-7728-6 (Paperback)
ISBN: 1-4107-7729-4 (Dust Jacket)

This book is printed on acid free paper.

1stBooks - rev. 07/26/03

DEDICATION

This book is dedicated to all those who love the work of the Lord as it pertains to His work for the dead.

CONTENTS

FOREWARD
by
Susan Easton Black
Professor of Church History and Doctrine
Brigham Young University

In *LDS Family History, the Easy Way*, Chuck Call shares insights learned from his ten years of teaching family history in local Sunday School classes. In *The Easy Way*, he provides just the right combination of doctrine, motivation, and "how-to" advice to delight the Latter-day Saint novice and expert genealogist alike.

His folksy writing style and clever narrative interpretation of family history makes this book a "page turner." Such quips as, "I have never, ever, heard of one of my dead ancestors refusing baptism," are guaranteed to be repeated more than once. This type of "pick me up" writing, even when discussing the "genealogist's brick wall," is refreshing. "Every pedigree line ends in one. But that is where the miracles happen." writes Call. He then shares his own miracle of visiting an ancestral graveyard and how ancestors buried there "became very real" to him and that they seemed to crowd upon him until he did their work at the Chicago Temple.

His inspired family stories aren't just "add-ons" to the text. They illustrate important principles like optimism, perseverance, and inspiration which are often associated with family history research. His anecdotes are also upbeat. "Just because grandma can't remember who *you* are," he confides, "doesn't mean she can't remember who her mother, father, sisters and brothers were."

But what will catch every reader's attention, again and again, is the central purpose of the text—(1) teach the doctrine of family history work, (2) teach how to create names for the temple, and (3) give the reader an idea of how to do easy research. The three-fold purpose is evident throughout the eight chapters in *The Easy Way*. From understanding doctrine to the beginning steps of genealogical research, the purpose is front and center.

The centrality of this approach assures readers of the importance of family history work and the binding power inherent in temple sealing ordinances. Quotations from Joseph Smith reinforce the purpose as the Prophet speaks of ordinances for the dead as "glad tidings of great joy" (Doctrine and Covenants 128:19). And as this joy gains a foothold with readers, there will

ix

emerge a renewed confidence in the readers' ability to discover anew family history—to discover how to become a Savior on Mt. Zion and "to do for others what they cannot do for themselves."

To further encourage readers to move forward with confidence, the author introduces helpful but easy research methods. A growing appreciation for computers, internet, FamilySearch, and E-mail will become evident. A reduced need for extensive genealogical travel and time spent in musty archives will be realized as readers better understand the value of technology and the important role of the stake family history centers. In those centers, family history research will become similar to working a "picture puzzle"—usually one piece doesn't tell the whole story but several pieces do.

If readers like solving puzzles, helpful clues for finding information on ancestry is awaiting them in *LDS Family History, the Easy Way*. The author encourages his readers to bless the lives of their ancestors and by so doing, bless their own lives by finding the joy awaiting them in family history.

"So what are you waiting for? Go make a difference!" says the author. And for those who have already made a difference, he says, "Keep it up! You're doing a great job!" And as for me, I add, "Chuck Call, you are doing a great job. Thank you for encouraging your readers to discover the happiness and joy awaiting them in *LDS Family History, The Easy Way*.

PREFACE

For ten years, I have been teaching family history in Sunday School classes. My syllabus was formed and reformed over that period of time, until our wonderful LDS Church programmers embedded some neat, online lessons within the "Help" menu of *Personal Ancestral File* (PAF). I then integrated those online lessons into *my* class lessons as homework for my students. It worked so well that I thought it was time to put the eight EASY lessons in print, so others could benefit from what I had learned.

This little book was written to take the mystery out of family history and to make it EASY — for new converts and "old timers" in the Church. It is also written so that *any teacher* can pick it up and teach eight family history lessons — EASILY. Additionally, it is written for the priesthood leader, who wants an EASY way to get a family history initiative started under his stewardship. And, lastly, it shows the EASY (and *only*) way to perfection — through family history and temple work.

My lessons were already in outline form on my Palm Pilot, so transferring those outlines to the word processor was no trouble. Then, all I had to do was "flesh them out" with the stories and doctrine I actually taught in class.

As I progressed through the book, chapter by chapter (and week by week, with my class), I found that it was EASY to give a copy of the next chapter to my assistant, Karena McGraw, and she could teach the class when I was out of town. She would feedback any errors in the manuscript, and I would make the corrections — even when I was out in California.

The manuscript has received many reviews by friends and family history associates. With their help, it's evolved quickly and easily, which is how things *should* happen — EASILY.

You may wonder why this little book has so many supplements at the end. Well, as the book developed, I began to realize that Church software and websites are changing so fast that if I put my "click-by-click" instructions in the body of the book, I would be forever reformatting it! By putting them in the supplements, whenever I need to update my instructions, all I have to do is update whatever supplement is affected.

I hope you enjoy reading and using this book, as much as I did putting it together.

Visit my website at
www.ldsfamilyhistorytheeasyway.com
to download an eBook

If you find errors in this book, please give me feedback at
cmcalljr@mybluelight.com

INTRODUCTION

As is written on the cover of this book, our Savior said His yoke was *easy*, not hard or difficult [Matt. 11:30]. The prophet Alma testified of the same thing in his instructions to his son, Helaman:

> O my son, do not let us be slothful because of *the easiness of the way*; for so was it with our fathers; for so was it prepared for them, that if they would look they might live; even so it is with us. The way is prepared, and if we will look we may live forever.[1]

Life in the "Savior's lane" is **not meant to be tough**! And **neither is family history work**. There *is* an EASY WAY to do <u>LDS</u> family history work. (I underline that, because there is a definite difference between LDS family history work and the genealogy "of men". <u>It's all in the goal</u>. Doing six generations of <u>ordinances</u> nowadays is a snap! You won't become a "genealogist", as <u>the world</u> sees it, but you *will* become a "savior on Mt. Zion".)

The Church is producing software and Internet helps for us, nearly at the speed of light! Well, there seems to be something new nearly <u>every week</u> anyway. And the temples have never been closer to us! Truly the Lord is making this work *easier* by the day.

This little book is my effort to help you see past what appears to most Saints to be "mountains" of research. I want to show you the EASY WAY to save your ancestors from the Spirit Prison. You can be a "Rambo" to your progenitors…and they'll love you for it…and they will bless your family's lives.

The goal is to: Do their ordinances for them

And it has never been easier in the history of this world.

IT'S EASY!!!
You won't even break a sweat.

---

[1] Alma 37:46, emphasis added.

I never have in 12 yrs!

But I *have* been instrumental in saving hundreds from prison.

Read on…

ACKNOWLEDGEMENTS

Most of all, I wish to publicly give thanks to my God for His help in bringing this book to you. I have felt His guidance and assistance throughout the entire process. He has taught me by incident and revelation, bringing me expert help in the daytime and intelligence by night. Without Him, it would never have come to be.

My heartfelt thanks also go out to the following people:

My loving and devoted wife, Kay, for her undying loyalty and the encouraging words, "You should write a book."

Karena McGraw and Christie Wall, for their encouragement after reading Chapter One. Had it not been for their enthusiasm about it, the project would have died at that point.

My brother, Mike, for his excellent review, advice and counsel.

Rita and Dale Millis, for their review of the rough draft and their calling it "good."

My son, Heath, Lance Morgan, and Mike & Kathy Ford, for their website and ebook expertise.

And last, but not in any way least, all of my past Sunday School students, who have been my "guinea pigs" over the years.

CHAPTER ONE
"Understanding The Doctrine"

How many times have we heard in meetings and conferences, a speaker, flaming with the fire of missionary work, quote the Prophet Joseph Smith:

> After all that has been said, *the greatest and most important duty* is to preach the Gospel.[2]

having heard not many weeks before someone else, equally aflame with the fire of family history work, quote the Prophet as saying:

> *The greatest responsibility* in this world that God has laid upon us is to seek after our dead.[3]

How can this be? How can *two* "folds" of the three-fold mission of the Church be *the* "greatest" or "most important" work that we can do? Is there a conflict here among the Prophet's statements?

Not at all.

It's merely a matter of "context". What were the circumstances of the statements? When were they made?

The first statement (regarding missionary work) was made on April 6, 1837. The Church was in its infancy. Of course, the most important work to be done was to spread the Gospel! Elijah had just restored the sealing power one year earlier! The Nauvoo Temple (the first to be built with a baptismal font for ordinance work) hadn't even been thought of, let alone built! Work for the dead didn't begin until sometime between 1840 and October 1841.

Joseph's statement about work for our dead wasn't given until Sunday, April 7, 1844...two months before his death. By that time, the Nauvoo Temple was well underway. Baptisms were already being held in the dedicated font area. The great era of work for the dead had begun!

---

[2] Roberts, Brigham H., *Documentary History of the Church*, Vol. 2, p. 478, emphasis added.
[3] Smith, Joseph Fielding, ed., *Teachings of the Prophet Joseph Smith*, pg. 356, emphasis added.

What greater missionary work can we do? I have never, ever, heard of one of my dead ancestors refusing baptism! In the October conference of 1893, President Lorenzo Snow said this:

> I believe, strongly too, that when the Gospel is preached to the spirits in prison, the success attending that preaching will be far greater than that attending the preaching of our Elders in this life. I believe there will be very few indeed of those spirits who will not gladly receive the Gospel when it is carried to them. The circumstances there will be *a thousand times more favorable*.[4]

STAY IN CONTEXT

Yes, staying in context is always important, lest we get carried away down a doctrinal path that may not be true. For example, if we read what the Savior said in the King James Version of Matthew 6:28-34,

> ...take no thought, saying, What shall we eat? or, What shall we drink? or, Wherewithal shall we be clothed?...,

we come to the conclusion that we shouldn't worry about where our next mouthful of food will come from. Sounds like a conflict with the Brethren's counsel to store a year's supply of food and clothing, doesn't it? That's because the Savior didn't say that *to the multitude*. He spoke those words to the Twelve, who were being trained for proclaiming the Gospel <u>without purse or scrip</u> (compare the Joseph Smith Translation of Matt. 6:1 & 7:1). Jesus taught the same things to the Twelve disciples in the Book of Mormon (3 Nephi 13:29)

WHY DO WE DO WORK <u>FOR</u> THE DEAD?

I want to emphasize the word "FOR" in the subtitle. That's what this chapter is all about. <u>Selfless service</u>. In *A Member's Guide to Temple and Family History Work* (a copy of which you should obtain from your bishop/branch president), Elder Boyd K. Packer declared:

> Once we have received [temple ordinances and covenants] for ourselves and for our families, *we are obligated* to provide

---

[4] Stuy, Brian H., ed., *Collected Discourses*, vol. 3, Burbank, California, and Woodland Hills, Utah: B.H.S. Publishing, 1987-1992, emphasis added.

these ordinances vicariously for our kindred dead…[Emphasis added]

Obligated? That's a pretty strong term! It means "compelled", "required", "duty-bound". Why? Why is there such a strong admonition for doing work for people we don't even know? (Well, at least, we don't *remember* them.)

It has to do with D&C 128:15. Therein, Joseph Smith (who was in hiding at the time) wrote the Saints in Nauvoo:

> And now my dearly beloved brethren and sisters, let me assure you that these are principles in relation to the dead…that cannot be lightly passed over, as pertaining to *our salvation*. For their salvation is necessary and essential to *our salvation*, as Paul says concerning the fathers — that they without us cannot be made perfect — *neither can we without our dead be made perfect.* [Emphasis added]

We've all heard this quote before. And then someone will come along who preaches that we can't be perfect anyway! Is that so? If we can't become perfect, why did the Savior <u>command</u> it in Matt. 5:48 and 3 Ne. 12:48? Ah! And in the comparison of those two verses lies the secret of the "principle of perfection". It is this powerful principle that I want to address now.

THE PRINCIPLE OF PERFECTION

We *can* become perfect, but it requires us to become like our Lord Jesus Christ. We must *become* perfect, like He did — the same way He did.

How did Jesus become perfect?

Compare those two verses again. What is the *context* of each? There you have the crux of the whole matter of perfection.

> He *became perfect* when He became our Savior — **by doing for us what we couldn't do for ourselves** — by atoning for our sins and breaking the bands of death.

The 93rd Section of the Doctrine & Covenants supports this principle by saying that "…he received not of the fulness at the first, but received grace for grace; And he received not of the fulness at first, but continued from grace to grace, until he received a fulness; And thus he was called the Son of

God, because he received not of the fulness at the first." As Jesus served others (i.e., gave "grace" by His miracles and teachings), he received grace from above—progressing, as we do, line upon line, precept upon precept, until He received a fullness of His Father's power and glory (i.e., <u>perfection</u>).

So Jesus became perfect **by doing for others what they couldn't do for themselves**.

And *that* is the "principle of perfection".

**We become perfect by doing for others what they can't do
for themselves**

That's how Jesus became our Savior, and it is how *we, too*, can become saviors—Saviors on Mt. Zion. In his book, *Decisions for Successful Living,* Harold B. Lee wrote:

> ...our Father...has provided a way for all members of his Church and Kingdom on the earth to be "saviors on Mt. Zion" by performing a vicarious work in behalf of those in the world of spirits...*that they could not perform for themselves.*[5]

STEPS ALONG THE PATHWAY TO PERFECTION

What are the steps to perfection? Well, we know that the first is:

<u>Faith</u> in the Lord Jesus Christ.
Then comes
<u>Repentance</u> for our sins,
followed by
<u>Baptism</u> by immersion for the remission of those sins
and
<u>Confirmation</u> as a member of the Church.
Hopefully, this is followed by
Reception of the <u>Gift of the Holy Ghost</u>,
<u>Priesthood</u> <u>ordination</u>,
and the temple ordinances of
<u>Endowment</u>
and
<u>Sealings</u> to parents and spouse.

---

[5] Lee, Harold B., *Decisions for Successful Living,* (1973), pg. 118, emphasis added.

Lastly, comes "enduring to the end" of our lives through an endless cycle of

Faith,
Hope,
and
Charity

And what is charity?

Why, it's *selfless* service, of course. And <u>that's what temple work for the dead is</u>:

Selfless Service, or in other words, **Charity**

Elder Henry B. Eyring, in a videotaped family history training session, told how those who have died can acquire faith and even a burning desire for baptism, but then they must wait...because the ordinances of baptism, confirmation, priesthood ordination, washing, anointing, endowment and sealings can only be done by flesh and blood people in earthly temples.

WHY THE ENDOWMENT IS SO IMPORTANT

Each ordinance is crucial to the salvation of the person for whom it is done. But the endowment has farther-reaching consequences than just for that person. Let me give you an example.

My father descends from Cyril and Sally Call. My mother descends from Elihu and Lola Allen. Each of these couples were the original "joiners of the Church" in their families, in the early days of the Church — the early 1830's. Each couple had burning testimonies that carried them through the Kirtland days, the mobbings at Far West, the escape to Illinois, where they lived within forty miles of each other — not in Nauvoo, but in Adams County.

Each of these couples felt the joys and sorrows of the Nauvoo period. Journals show that they each passed through the gloomy days after the Martyrdom of Joseph and Hyrum. Each went through the mobbings of the following months, and each sought shelter in Nauvoo. Each lost small children due to these persecutions.

Each couple slogged through the mud of the 1846 Iowa exodus, and each went through the struggles of the Winter Quarters period. The Allens, strong in the faith, were so anxious to "follow the Brethren" to the mountains, that they were enlisted in the first big company of pioneers to go West. They arrived in the Salt Lake Valley in September of 1847, but, by the time the Calls arrived in 1850, Lola and her daughter, Lola Elizabeth, were already dead. Elihu followed them to the grave in October, and the remainder of the family was scattered. Many were lost to apostasy.

The result? While the Call descendants are numerous and quite active in the Church, the active descendants from this Allen couple are but few. My mother is a *convert* to the Church! The <u>only</u> member in her family!

How could that happen? What made the difference? Have you guessed it? The endowment. The Calls received their endowments in the Nauvoo Temple, before they left in 1846. The Allens did not. It's that simple, and it's that tragic.

The power of the endowment on your posterity is tremendous! And that power is multiplied each time one of your ancestors receives their endowment, because the endowment is for them *and their posterity*, as shown in Fig. 1. Is this what the Lord means in the Doctrine & Covenants, when He says, "I will bless him with a *multiplicity* of blessings"?

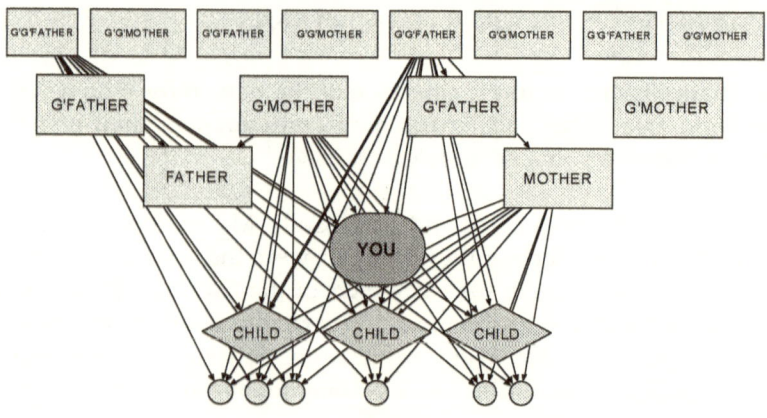

Fig. 1
The Effect Of The Endowment

The power of the endowment extends to your children and grandchildren both <u>physically</u> and <u>spiritually</u>. It protects them from both seen and unseen dangers and from the powers of darkness. Do you see why your salvation is dependent upon doing these ordinances for your ancestors? The power of each new endowment strengthens you and your posterity against the onslaughts of Satan's armies, who are very real.

And just as Satan is very much alive, so are your ancestors. **They live**, because of our beloved Savior. They wait only to be released from prison by baptism, to join His Church by confirmation, to receive the priesthood and the promises of the initiatory ordinances, to receive the power of the endowment, and to be sealed to you by a "welding link", which I will address in the next chapter.

As Elder Eyring so eloquently put it, "Even Joseph, and Hyrum, and Brigham *must wait on us*" to do the baptisms for those whom they have loved and converted in the Spirit Prison.

So what are you waiting for?
Go make a difference!
IT'S EASY!!!
Read on...

CHAPTER TWO
"Getting Started"

We learned in the preceding chapter that the "principle of perfection" is:

**You are perfected by doing for others what they can't do for
themselves**

In order to become like the Savior is, we must do what He did — be saviors —
on Mt. Zion.

We also learned the importance and power of the endowment ordinance —
that it extends to the posterity of the person receiving the endowment. We
also learned that the more ancestors you have who have received the
endowment, the more *you* and *your posterity* are blessed and protected from
the powers of darkness.

It's similar to the principle behind temple prayer rolls. The more people
that are involved in a prayer, the stronger the faith, and the more power for
a miracle. Let me give you an example.

When our oldest son was on his mission in Tokyo, he started having some
sort of problem, the details of which I no longer remember. I hasten to say
that it was not anything sinful. However, it was affecting his attitude as a
missionary. His mother and I became quite concerned and felt something
had to be done.

Accordingly, we called his mission president, who was aware of the
problem. We set a date that we would fast and pray with him, and, at the
end of the day, he would call our son and ask him to come in for an
interview to discuss the problem.

We also asked the president to put our son's name on the prayer roll of the
Tokyo temple. We did the same at the Chicago temple. Then we called all
of our relatives, on both sides of the family, and asked them to fast and pray
with us on that day — and to put his name on the prayer rolls of their
temples. So when the actual fast day arrived, we had quite a few people all
over the world fasting and praying for a miracle to happen in our son's life.
His name was on the prayer rolls of at least six or seven temples!

And it worked!

The faith and prayers of so many turned the key, a miracle occurred, and a young man's mission and future were saved. Such is the power of the endowment when it is given to hundreds of your ancestors.

THE DEAD ARE VERY MUCH ALIVE

In the last chapter, we learned how family history and temple work perfects *us*...the living. But, how are the dead perfected? The scripture clearly states that this work is also to perfect them.

> And now my dearly beloved brethren and sisters, let me assure you that these are principles in relation to the dead...that cannot be lightly passed over, as pertaining to our salvation. For *their salvation* is necessary and essential to our salvation, as Paul says concerning the fathers — that *they without us cannot be made perfect* — neither can we without our dead be made perfect.[6]

How is it done? By the same "principle of perfection." The dead who receive baptism, membership in the Church, ordination to the Priesthood, and the endowment, then have the power within them to **do for others what they can't do for themselves**.

How? One thing they can do is to serve missions to others in the Spirit Prison. Another thing they can do is to help their descendants find earthly records that can free more ancestors from prison. But they can also do good for those who have been the instrument by which they were made free — those who have done the family history and temple work.

"I AM NOW WHERE I CAN DO YOU GOOD"

My third great-granduncle, Anson Call, recorded in his journal a fascinating vision or dream, which he had following the martyrdom of Joseph and Hyrum Smith.

> On Monday the bodies of Joseph and Hyrum were taken to Joseph's mansion and were exhibited to all who wished to see them. I accordingly took my family and took a view of the bodies of the Prophet and Patriarch and returned home with my family. Sleep and the desire of food had left my

---

[6] D&C 128:15, emphasis added.

body. I shall not attempt to describe my feelings. What was to be done, I knew not. I cried mightily unto the Lord that I might know what to do. The third night I had a dream, or a vision, which I will relate.

I was traveling by myself in a lonely place till I came to a new field about 3 acres in size. I discovered in the center of the field a nice block house. I went to the door of the house. I discovered Joseph in the house standing in the middle of the floor. I sprang and clinched him by the hand. I threw my arms around him, and kissed him, and said, "Joseph, I thought you was dead." He said, "I am." I said, "This is certainly Joseph." He said, "Yes, it is Joseph. Take your seat and I will tell you all about it." I seated myself and then discovered I was in a congregation of saints whom I was acquainted with. Joseph then said, "Brethren, I have been killed in Carthage jail, and it will not make any difference with you, if you do as you are told. I shall continue to govern and control this kingdom as I have hitherto done. The keys of this kingdom were committed to me. I hold them and shall continue to hold them, worlds without end. I am dead, and I am out of the power of my enemies. *I am now where I can do you good.* Be no longer troubled. Be faithful, be diligent, do as you are told, and you shall see the salvation of God."

I then discovered myself sitting in the bed. I spoke to my wife and told her what I had seen and heard and let her heart be comforted, for we should yet see good days and prevail over our enemies.[7]

The part that I want to draw your attention to is that Joseph, being dead, was now in a position to do more good. This is also born out in a quote from Joseph F. Smith:

Nature of Ministering Angels

We are told by the Prophet Joseph Smith, that "there are no angels who minister to this earth but those who do belong or have belonged to it." Hence, when messengers are sent to minister to the inhabitants of this earth, they are not

---

[7] Call, Ethan L. & Christine Shaffer, *Journal of Anson Call*, pg. 29, emphasis added.

strangers, but *from the ranks of our kindred*, friends, and fellow-beings and fellow-servants.[8]

My grandmother was a good, kind woman. She was always looking after my welfare. She was the "mediator" between my Dad and I. She made sure I was baptized at eight years old (even though we weren't active in the Church). She made sure I got my driver's license and my car. She was always looking out for me.

One summer, my wife and I were asked to give four class presentations in a youth conference on any subjects we chose. We decided to split them up. I would do two, and my wife would do two. I had the first one pretty well figured out, but I was stumped on a topic for the second one. At the time, I was working for the Church, and one day I was traveling across the state of Missouri, listening to a Paul Dunn tape entitled "Your Cheering Section". On that tape, Brother Dunn said that we each have a cheering section on *both* sides of the veil.

I considered this as a topic to use, and as I did, a thought came into my mind about a time just before my 35th birthday, when my life took a dramatic turn for the better. It was a pivotal point in my life. I was called into a bishopric and had to do some deep soul-searching and covenanting with the Lord before I felt right about accepting the calling.

As I mused upon that event and the Paul Dunn tape, I wondered if my grandmother had been in *my* "cheering section" at that time? I was almost sure that she was not alive at the time.

When I got to my motel that night, I called my Dad and asked him when his mother had died. He couldn't remember exactly, but thought it was sometime in 1977. That was the year! I asked him if he could pin the date down? My mother got out her family history and gave me the date — 24 April 1977 — *exactly one week before I was ordained a High Priest and set apart as a counselor in the bishopric!* Was she in my cheering section? You bet! And the Spirit bore strong testimony to me that night, and still does, even as I write this, that she was the moving force behind that calling, which changed my life. Yes, the dead *are very much* alive.

When Jesus struggled in the Garden of Gethsemane for the sins of mankind, sweating "...as it were great drops of blood falling down to the

---

[8] Smith, Joseph F., *Gospel Doctrine*, Salt Lake City: Deseret Book, 1939, pg. 437, emphasis added.

ground,...there appeared an angel unto him from heaven, strengthening him." (Luke 22:43-44) Who was that angel? Elder Bruce R. McConkie reasoned that it surely would have been Adam. I don't think so. I believe it probably was His grandmother. Jesus *did* have a grandmother, you know, and I'm sure she was even kinder than mine.

WHY ARE SEALINGS SO IMPORTANT?
WHAT DO THEY DO?

Brigham Young was a lot like Father Abraham—he knew an awful lot about astronomy and, more especially, about the creation of this earth. Something he said in 1874 about the Fall of Adam is significant in regards to the sealing ordinances:

> This earth is our home, it was framed expressly for the habitation of those who are faithful to God, and who prove themselves worthy to inherit the earth when the Lord shall have sanctified, purified and glorified it and brought it back into his presence, *from which it fell far into space.* Ask the astronomer how far we are from the nearest of those heavenly bodies that are called the fixed stars. Can he count the miles? It would be a task for him to tell us the distance. *When the earth was framed and brought into existence and man was placed upon it, it was near the throne of our Father in heaven.* And when man fell—though that was designed in the economy, there was nothing about it mysterious or unknown to the Gods, they understood it all, it was all planned—*but when man fell, the earth fell into space, and took up its abode in this planetary system, and the sun became our light.* When the Lord said—"Let there be light," there was light, for the earth was brought near the sun that it might reflect upon it so as to give us light by day, and the moon to give us light by night. This is the glory the earth came from, and when it is glorified it will return again unto the presence of the Father, and it will dwell there, and these intelligent beings that I am looking at, if they live worthy of it, will dwell upon this earth.[9]

We're a lllllooooooooonnnnnnnnnngggggggg way from home, aren't we? Our Heavenly Home. We're way out here in space. How do the angels in

---

[9] *Journal of Discourses*, vol. 17, London, Latter-day Saints' Book Depot, 1854-1886, pg. 144, emphasis added.

Heaven keep track of the "billions and billions" of us, who *have* lived, *now* live, and *yet* will live, upon this earth? By magic? Of course not! By the sealing power of the Holy Priesthood. When Jesus promised that power to Peter, He said,

> I will give unto thee the keys of the kingdom of heaven: and whatsoever thou shalt bind on earth shall be bound in heaven: and whatsoever thou shalt loose on earth shall be loosed in heaven.[10]

Joseph Smith gave a more detailed explanation:

> Now, the nature of this ordinance consists in *the power of the priesthood*, by the revelation of Jesus Christ, wherein it is granted that whatsoever you bind on earth shall be bound in heaven, and whatsoever you loose on earth shall be loosed in heaven. Or, in other words, taking a different view of the translation, *whatsoever you record on earth shall be recorded in heaven*, and whatsoever you do not record on earth shall not be recorded in heaven…[11]

In a nutshell, the sealing power of the Priesthood is the power to record ordinances and relationships between persons *on this earth*, and at the same time, record them also *in heaven*,

So they won't be lost in the immensity of space.

This is the "…welding link…between the fathers and the children…" spoken of by Joseph in verse 18. A power that reaches out through space and welds together "…dispensations, and keys, and powers, and glories…" and *families*. Do you now see why sealings are so important? Without them, this whole earth would be "utterly wasted".

THE GOAL OF LDS FAMILY HISTORY WORK

Have you ever seen the 3x5 blue, pink and yellow *family file* slips that are used in our temples to do ordinances for our own personal ancestors? They're a lot different than the ones you are given from the *temple file*. On the slip, is all of the information necessary to do *all of the ordinances* shown

---

[10] Matt. 16:19
[11] D&C 128:8, emphasis added.

14

on the slip—for that person.   And *that* is the goal of LDS family history work:

You want to provide complete enough information to identify "...your ancestor *uniquely* so that he or she cannot be confused with another person"[12], *so you can produce a family file slip for them.*

**The goal is to do the ordinances for them.
Never lose sight of the goal!!!**

On page 10 of *A Member's Guide to Temple and Family History Work*, the minimum requirements for baptism, endowment and sealings is given in chart form.   I repeat them here to show you how minimal they really are, and some of this information may even be estimated!

FOR BAPTISM AND ENDOWMENT
- Name
- Sex
- Event date (for example, a birth date)
- Event place (for example, a birthplace)

FOR SEALING TO PARENTS
- Information under "For Baptism and Endowment"
- First or last name of the father

FOR SEALING TO SPOUSE
- Name of the husband
- Marriage date
- Marriage place

That information is *all* that is required to do the ordinances. **You don't have to fill in the endless blanks of a family group sheet**. That is *not* the goal.

Don't get me wrong.   The more information you can provide for a person, the more certain you will be of that person's relationships to your family, and it will make further research easier for you and other researchers, *but it's not the goal.* The goal is to save souls. The goal is to give them the power to save others.   The goal is to bring "glad tidings of great joy" to the captive spirits by releasing them from prison.

---

[12] *A Member's Guide to Temple and Family History Work,* (1993) pg. 10, emphasis added.

Take the time *now* to sit back and imagine the joy and rejoicing that fills the hearts of those who are privileged to see the prison gates swing open for them.

> How beautiful upon the mountains are the feet of him that bringeth good tidings...[13]

> Now, what do we hear in the gospel which we have received? A voice of gladness! A voice of mercy from heaven; and a voice of truth out of the earth; glad tidings for the dead; *a voice of gladness for the living and the dead*; <u>glad tidings of great joy</u>. How beautiful upon the mountains are the feet of those that bring glad tidings of good things, and that say unto Zion: Behold, thy God reigneth! As the dews of Carmel, so shall the knowledge of God descend upon them![14]

> While this vast multitude waited and conversed, rejoicing in the hour of their deliverance from the chains of death, the Son of God appeared, declaring liberty to the captives who had been faithful...[15]

Can you imagine how they feel about those who have made their freedom possible? How does a POW feel towards those who finally set him free? *Can you imagine the love your ancestors will feel towards you when you open the gate from darkness into eternal light?* There won't be enough that they feel they can do for you, to repay you for your efforts. And you can free an *army* of them! An *army* willing to help you in any way they can! The Holy Ghost makes you want to help them, and He also makes them want to help you.

When I kneel down and pray for angels to protect my son or daughter, whom do you think God will send? Your ancestors or mine?

EASIER WAYS

The Lord has really made family history work easier than it's ever been before. With computers, the Internet, the FamilySearch website and email, what used to take years of traveling to county seats and searching through dusty records, is now able to be done in weeks on the Web, by collaborating

---

[13] Isaiah 52:7
[14] D&C 128:19, emphasis added.
[15] D&C 138:18

with cousins all over the country. Even old photos can now be sent by email. What used to mean an expensive trip to Salt Lake City and the Family History Library, now can be accomplished in your own Church meetinghouse.

Even the differences between LDS and non-LDS methods of doing genealogy make obtaining information easier. You see, LDS genealogy is "ancestor-driven", while non-LDS genealogy is usually "descendant-driven". This creates what I like to think of as lots of giant "catcher's mitts" out there. I'll get into these in more detail in a later chapter.

HOW TO GET STARTED

Go to your nearest LDS Family History Center. There may be one right in your own meetinghouse. Get a copy of the brochure, *How Do I Start My Family History?* It's FREE!!!

Quickly pencil into the pedigree chart all that *you* know, back as far as you can. Once you have all you can *easily* put on paper, download the Church's FREE *Personal Ancestral File* software (PAF) from the Internet site, www.familysearch.org, using my instructions in Supplement A. Then, do "Help Lesson #1" from within the PAF program by following the instructions in Supplement B.

Be sure you always make *two* backup floppy disks at the end of the lessons, and store them in separate locations. This is to protect you from losing all of your hard work. Did I say "hard"? Shame on me! This is EASY. Have you strained a muscle or broken a sweat yet? Of course not!

The Savior said it would be EASY! Believe Him.

Read on...

CHAPTER THREE
"Getting Help"

In the last chapter, I explained how Joseph F. Smith had written that the angels who minister to *us* come "...from the ranks of our kindred..." They are familiar with us, and we with them. Let me tell you of *just one* of the ways in which your dead can help you.

For years, I used to read the "Family History Moments" on the back cover of the *Church News*. I just *knew* that those people were only imagining those miracles happening to them. And then, in 1991, when I started doing my own family history, those miracles started happening to *ME*!

"End-of-line individuals" are always miracle makers. Do you know what that term means? End-of-line individuals? It's those folks on your pedigree chart who are way off to the right, but they're as far as you can go on their line. They're at the end of a pedigree line, and you just can't seem to get beyond them. You've searched and researched, but there just isn't any information easily available to you which allows you to push back another generation. You've come to what genealogists call a "brick wall". Every pedigree line ends in one. But that's where the miracles happen!

Let me tell you about one. In 1992, just after getting started with my own genealogy, I was spending a night in a motel in West Frankfort, Illinois. I was there on some Church business I had to attend to the next day. So I thought I'd go to the public library, just to see what they had in the way of family histories. I had previously determined that a grandmother and grandfather had married in that area in 1826.

Well, I met some people who knew enough to steer me to information that allowed me to do the temple work for these "end-of-line individuals". In 1996, my family got the opportunity to actually move to West Frankfort! This is 3-4 years *after* their work had been done.

I visited the county courthouse to see the actual record of that couple's marriage (Martin Woolley to Nancy Neal). A worker at the courthouse who was helping me said, "Well, I know a Nancy Neal. She used to work here!" So I got Nancy's phone number from the worker, and called her. She lived right there in West Frankfort! She invited me over to visit with her and her daughter. They were amazed! I was a descendant of a Nancy Neal that no one else in the family had been able to trace. They knew she and Martin had left the area in 1847 and gone to Missouri, but that was as far as they had been able to trace them, let alone their descendants.

I was invited to attend the annual "Neal Family Reunion" held at the Neal Cemetery! There I met all my Neal cousins from all over the United States! And I made further connections and learned of a new book that had just been published by one cousin, which gave me enough information to do the work for another four generations of Neals! In that book, it told who my Nancy Neal's mother was. Nancy Jordan. But the author of the book couldn't trace the Jordans beyond her father, Thomas Jordan.

So we (my extended family) did the Neal's work for them. Now, remember, this is in 1996. By 1999, the author of the Neal book had completed a similar book for the Jordans, adding four more generations on that line!

How do such miracles take place? Well, I think it's this way. Who knows better where family records are, than the persons who lived through those recorded events? I feel that once those "end-of-line individuals" are released from prison, they are able to bring the missing records to light. This is just one way in which we receive help from the dead.

HELP FROM <u>THIS</u> SIDE OF THE VEIL

Now that you've completed "Help Lesson #1" in your *Personal Ancestral File* (PAF), you should have copied the names, dates, and places from your paper pedigree (*How Do I Start My Family History?*) into your computer pedigree. Now's the time to get the rest of your family involved. After all, did you think that *you* had to do all the work? No! Keep it simple.

"Many hands make light work."

Now contact your relatives and get their input. Remember, just because Grandma can't remember who *you* are, doesn't mean she can't remember who her mother, father, sisters and brothers were. Those are some of the only correct memories that old folks have. Never discount them. They love to talk about the times when they were young.

Not everyone has a gift for family history research. Some people revel in the "picture puzzle" detective work that is required. Others just cringe to think of it—and shrink from it altogether. But **all** *can* do something. All can be "saviors on Mt. Zion" in their own unique way. If a member of your family doesn't like to do research, they *can* type names into a computer, or they can do baptisms, confirmations, or endowments. Young and old can always do *something* to push the work along. Everyone's circumstances are different, but all *can* do something.

Collaborate with other family members. If you're a new convert to the Church and are the only member in your family, be careful. What you say, or how you say it, can close the doors to much information that would normally be freely given to anyone other than a "Mormon". It's sad, but it's true. It's amazing how closed-mouthed people get when they find out you're a member of the Church. So be careful not to say too much. Although we are told to "let our light so shine" by being missionaries, it seems that if we do it with non-member genealogists, we could jeopardize the freedom of hundreds of our ancestors who will *gladly* accept the truth. **After** you have gleaned all of the information you can from a resource, **then** is the time to give them a Book of Mormon.

If you have other active members in your family, there's a good chance that one of them is taking this work seriously and already has a head start. Collaborate with them. Have them give you their information as a starting point. Get it into your computer and work together on it. You may live in a part of the country where you have access to records that they've wanted to see.

In my ward, I have a good friend, Les Scannell, who is the only member of the Church in his family. But his sister (a non-member) has done an excellent job in genealogical research for their family. She has given him a 1 ½" thick binder of their family lines. I've helped Les a little by typing some of those lines into his PAF file. After we get back a few generations on a line, we look in the Church's *Ancestral File* to see what will match up with his ancestors. Every time, we have found that his ancestry runs back to patriots in early colonial America! Les could easily become a member the *Sons of the American Revolution (SAR)*.

STEPS TO THE TEMPLE

With all of the information you will be putting into the computer and gleaning from other family members, it might get a little confusing. Figure 2 can help you keep your perspective.

DON'T lose sight of the GOAL!!

To provide temple ordinances

Fig. 2
Steps in taking Ancestors to the Temple

YOUR MANUAL

A Member's Guide to Temple and Family History Work is a wonderful manual...and wonderfully *thin*, too! READ IT!!! **Cover to cover.** In its pages you will discover the "whys" for doing things the way we do. Read carefully pages 10 and 11. The information there is critical.

It's not necessary to "fill in all the blanks", but you *do* have to at least give all the minimum information. It's critical that you use proper formatting. Why? Because when you want to send your information to other members of the family, you will want to use what is called, in the genealogical "industry", **GEDcom files**. This is a computer file format that was originally invented by the Church to allow sharing of genealogical data. It stands for "Genealogical Exchange Data Communications File" and is simply called a "GED" by most of us. The Church set the standard for this file over a decade ago, and the *world* still uses it today!

Continue putting your information into your PAF file by going to Supplement C and following my instructions for "Help Lesson #2". Add pedigree as far back as you can go, and then add what you can obtain from other family members. Keep your *Member's Guide* close at hand while you do, making marginal notes and highlighting where you need to. **Be sure and read it TODAY!**

CHAPTER FOUR
"Movin' On"

In preceding chapters, I've told you that your goal in doing family history work is to do the ordinance work for "Ancestors". Just who are your "ancestors"? Are they your cousins? Are they your uncles and aunts? How about their spouses?

Elder Packer has told us that "...we are obligated to provide these ordinances vicariously for our *kindred dead...*"[16] Who are our "kindred dead"?

Who are you *really* responsible for?

The answer lies in your pedigree chart. For each married couple on your pedigree chart, there is a "family group", made up of a father, a mother, and their children. **You are responsible—obligated—to provide the temple ordinances for all the *family groups* on your pedigree chart**. Doing the math for six generations (beginning with your parents, then your grandparents, etc.) equals 31 family groups. (That's not counting your own family group with you as a parent)

If it's easy enough information to obtain, I like to do the children's spouses, too. But it's not required of *you*. If their children's information falls easily into my hands, I do that, too. But it's not required.

THE PURPOSE OF THIS BOOK

My purpose in writing this book is to (1) teach you the doctrine, (2) teach you how to create names for the temple, and (3) give you a smattering on how to do the EASY research. By the time you finish the book, you should be quite comfortable with PAF, and you should have enough experience under your belt to keep going on your own.

THE WONDERS OF PAF

At the time of this writing, PAF version 5.2.18 is the latest version available, and it is the best ever! There is no other program on the market that compares with it. And it's FREE!!! Plus it's designed with LDS family

---

[16] *A Member's Guide to Temple and Family History Work*, pg. i, emphasis added.

history in mind. It's set up to work with the newest version of TempleReady for Windows, so that the two programs "talk" to each other! That's a first!

Uh, oh! Now you want to know what TempleReady is. Well, it's the Church's computer program, available only at LDS Family History Centers, which clears names for the temple ordinances. But I'll get into that in a later chapter.

Each screen (or window) in PAF has a HELP button, plus a ? button. That's a nice feature in and of itself. All you have to do is click the ? and then click on the area of the screen that you have a question about. Bingo! There's the answer! Cool, huh?

The "Ditto" feature for places is great, but sometimes the list just gets too unwieldy, so you may have to "prune" it after a while. You can do that by double-clicking on any name, clicking on a **blank** "Place" box in the "Edit Individual" dialogue screen, then clicking on the down arrow at the right-hand end of the "Place" box, highlighting a place name in the drop-down pick list and pressing the "Delete" button on your keyboard. **Be careful**, though. **Be sure you always do this in a blank "Place" box**. If you don't, you will most probably also delete a good place name!

By this time, you should have learned all about changing the order of children and spouses. That's an important thing to know, because keeping the children and spouses in chronological order is just good, genealogical practice. It is assumed by the genealogy community, throughout the world, that children are always in their proper birth order and multiple spouses are in proper chronological order.

FORMATTING

Which brings us to the topic of "formatting". There's just no substitute for good formatting and no excuse for bad formatting. I can't tell you how many times I have received GED files from people who seemed to know what they were doing, only to find that their place name formatting was all wrong! Most of the time, they get the date format correct, although I've even seen some of those that were "mavericks". When you're collaborating with other people, it's just good etiquette to use the proper formatting. Again, I refer you to pages 10 and 11 of your *Member's Guide* for the proper formats to use.

I used to refuse to spell out states and put in "USA" as the country, but with the "Ditto" feature, I now have no excuse. All I have to do is type it once.

DOCUMENTING YOUR WORK (STATING YOUR EVIDENCE)

One of the benefits of doing this work is that you get to share it with others. If you have properly backed up your data with good references to your sources for birth, marriage, and death information, other people can use your information with confidence. They can then help others by combining their information with yours and deducing other important evidence, which may then open new doors to you. So be sure you always **document, DOCUMENT, <u>DOCUMENT</u>!!!** It pays big dividends.

Document your sources *as you enter information.* If you think you'll go back and do it later, you won't. You'll forget. <u>*The best time to do it is at the time that you enter the information.*</u> Sorry, but that's just the way it is! Two good books on proper formatting of your sources are *Evidence! Citation & Analysis for the Family Historian,* by Mills, and *Cite Your Sources,* by Lackey, but a high school English text may also be a good place to obtain guidance.

TYPES OF SOURCES

There are many types of sources of information. We've been dealing mainly with two, up to this point. Stuff that you remember or know to be true, and stuff that your family members remember or know to be true. To some genealogists, this is "secondary" information. "Primary" information is documentation such as birth, marriage, and death certificates. To them, a family bible is a "secondary source", because it wasn't documented *at the time of the actual event.* That is the definition of a "primary source". It is documentation done at the actual time of the event (i.e., birth, marriage, death, etc). However, most of the information you have gathered to this point has been "word of mouth" or from the personal records of someone you know and trust. For LDS family history, that is perfectly acceptable.

Remember, you want to provide complete enough information to identify "…your ancestor *uniquely* so that he or she cannot be confused with another person" [17], *so you can produce a family file slip for them.* That is the standard, the criteria, for your work. You don't have to have "primary" sources, **as long as you *know* the information is good and true.**

---

[17] *A Member's Guide to Temple and Family History Work,* pg. 10, emphasis added.

So when you are using the information you have gleaned from Grandma Jones, your source should be "Personal records [or memory] of Grandma Jones". And *where* are those records? Well, that brings us to the subject of repositories.

REPOSITORIES

Each source or document "reposes", or *resides*, or is de*posit*ed, somewhere. That physical place is called a "repository", and you should always give that information as completely as you feel it needs to be—taking into account the privacy and safety of those people who have given you the information. Perhaps it would not be safe for Grandma Jones, if you were to give out her phone number and address. But you can probably safely give her name and the city and state where she lives.

ACTUAL TEXT

Sometimes it's beneficial to copy the text of a document, word for word, because you may not have enough additional information to make sense of the entire document at that time, but subsequent information may come into your hands which will make the document completely clear to you. If you have the entire text in your source, you can go back to it at any time, and PAF allows you to do that.

IMAGES

Oh, boy! Isn't cyberspace grand? PAF allows you to attach images (scanned documents, photos, etc.) as documentation for your sources. So, if you have your scanner with you when you interview Grandma Jones, and she keeps referring to her family bible, you can scan those pages into your computer and attach the image to your source. What a country! What a time to be alive!

NOTES

Do not put sources in the "Notes" areas, although it is acceptable to put some notes in the Source areas. Suppose you just can't quite figure out where so-and-so was when he married such-and-such. You have a pretty good idea, but you just can't prove it. Put your thoughts down in the "Notes" for that person. Then when you go back to it, months later, you'll be able to "pick up the thread" of your thoughts again, without missing a beat.

If you have thoughts or comments that refer to a specific source, you should type those into the "Comments" box of the "Edit Source" screen.

HELP LESSON #3

Now's the time to do Help Lesson #3, as shown in Supplement C, and, by all means, if you haven't completely read your *Member's Guide*, again, I *urge* you to DO IT NOW.

Ain't this fun?

CHAPTER FIVE
"Downloading from the Internet"

By this time, you should have a pretty fair pedigree in your computer. Now's the time to see if you can save yourself some work. You can do that by searching the Church's Internet site at www.familysearch.org. Start by using your great grandparents' names. See if they can be found in the *Ancestral File* or the *Pedigree Resource File*. If they show up there, then someone has submitted them to the Church for sharing, and you may download any pedigree you find there.

ANCESTRAL FILE

Back in the middle 1970's, *before the advent of the desktop computer*, members of the Church were asked to submit family group sheets for the first four generations of their pedigree charts. A few years later, we were asked to submit *six* generations. *Little did we know*, back then, how that information was going to be used. But the pedigrees of those who were faithful and obedient became the nucleus of the Church's *Ancestral File* database, which began to appear on the first CDs in local Family History Centers across the nation in 1991. From that nucleus of pedigrees, has grown the present day *FamilySearch* website, which grows more breathtaking by the day! That just goes to show you what wondrous things are brought about through faith and obedience.

You already know about *Personal Ancestral File*. Well, *Ancestral File* preceded it, and it is **just a piece** of the entire *FamilySearch* program. *Ancestral File* is the **Church's** database, and PAF is *your* **personal** database. Do you have any relatives who may have contributed information to the *Ancestral File*? If so, you need to go to www.familysearch.org and see if you can download that information onto your computer's desktop (see Supplement D for instructions on how to do that).

If you don't have the Internet available, you can go to your nearest LDS Family History Center and search their *Ancestral File* CD set. You can download anything you find onto a floppy diskette to be imported into your PAF. The staff at the Family History Center can show you how that's done, or direct you to someone who can.

> **NOTE**: This is something that is so important that I've inserted it here *and in* **Supplement D.** What you see is what

you get (WYSIWYG). The Internet *Ancestral File* is different from the CD version, which you will find at your local Family History Center. The Internet version is WYSIWYG (said "whizzy-wig," in computer jargon). <u>In other words, you can only download the "chunk" of a pedigree that you actually **see** on your computer screen!</u> If the pedigree you're looking at has "right arrows" at the ends of some of its family lines, then you can click on those arrows, one-at-a-time, to bring to your screen additional "chunks". You can then download those "chunks" as separate files. It's slower, but it's a lot more up-to-date information. The CDs at your FHC could be two years old. The Internet site is as fresh as it gets!

IMPORTING INFORMATION INTO YOUR PAF FILE

After you have your download file stored either on your computer desktop or on a floppy disk, then you can import that information into your PAF file. You do that by the following steps:

- ❏ Click the **PAF 5.2** icon on your desktop
    - o This should open the most recent PAF file that you've been working on, which should be yours.
    - o If it does not open your file:
        - ▪ Click on the **File** menu
        - ▪ Select your file at the bottom of the drop-down menu
        - ▪ This should open your file
- ❏ Click on the **File** menu
- ❏ Click on **Import**
- ❏ In the **Look in:** box, click the down arrow
- ❏ Select either **Desktop** or **3 ½ Floppy (A:)** from the pick list, depending on where you downloaded your GED
- ❏ Select the GED file you downloaded (the one with your ancestor's surname)
- ❏ Click the **Import** button
- ❏ Leave the boxes checked as they are, and click **OK**
- ❏ After the import is completed, click **Yes** and review the Help information regarding imports.
- ❏ Close the Help window

You now have your ancestor's pedigree "tree" in your PAF file. However, the two pedigree "trees" are still separate. Let me explain this in another way.

Suppose you had a manila file folder, which you have named "PAF". Then suppose you put into that folder a paper copy of your pedigree. Then you print out your ancestor's pedigree on a separate piece of paper, and you put that into the folder, too. Now you have two separate "paper" pedigrees in the same "PAF" file folder. That's what you have, electronically, in your PAF file in your computer — two "electronic" pedigrees in an "electronic" file.

For a visual understanding, see Fig. 3. If GGRANDPA is your ancestor, you see that he is in your PAF twice. Click on the **Individual** tab and look for your ancestor's name. You should see it in the listing, twice, with two different **RIN** numbers.

In order to attach your ancestor's pedigree to yours, so that it becomes an extension to yours, you will have to do a **Match/Merge**, which will be explained in the next chapter. In the meantime, if you haven't yet read the *Member's Guide*, now's the time to do it.

READ THE MANUAL >>> TODAY

And now's the time to do **Help Lesson #5**. We'll leave Help Lesson #4 'til later! I think it fits in better at that time. At the end of Help Lesson #5, you will be introduced to TempleReady, but don't do anything with TempleReady yet. We'll discuss that in Chapter Seven.

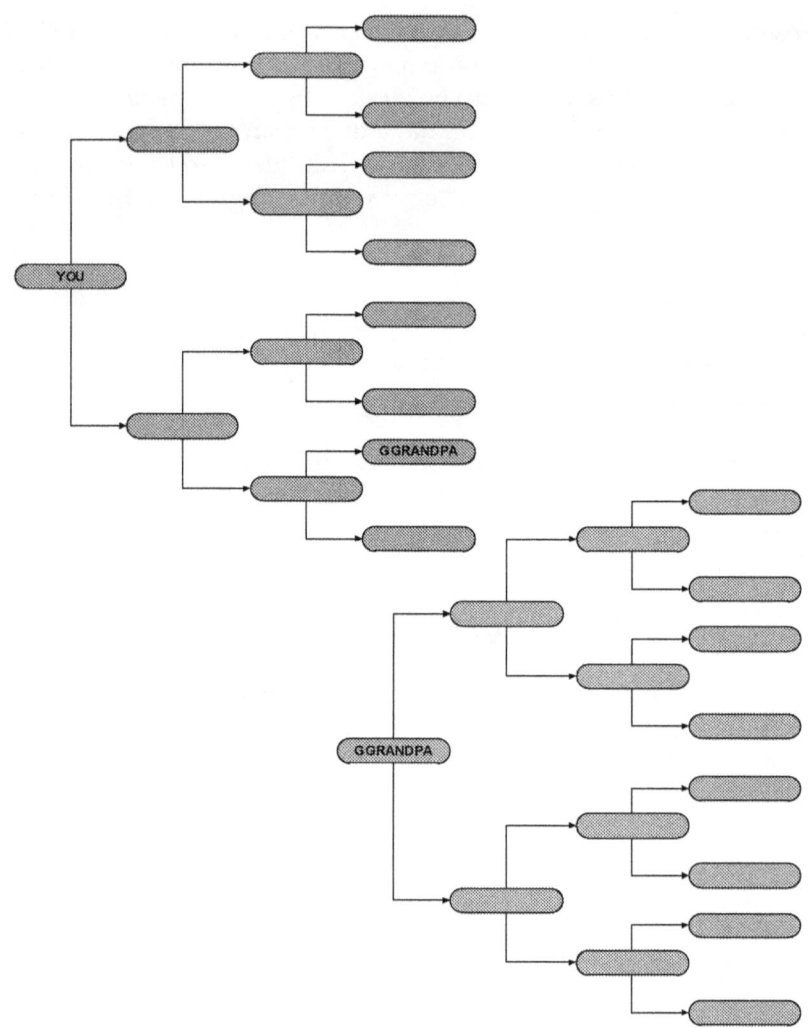

Fig. 3
Separate pedigrees in PAF

CHAPTER SIX
"Merging Info"

Sometimes, we may have had a dead ancestor who was such a reprobate during his earth life that we may be tempted to neglect doing his temple work. We just "know" that it would be a waste of time, don't we?

**Don't fall prey to such a temptation.
That's what Satan wants you to do.**

We don't have the right to judge anyone as to whether or not they are worthy of the ordinances. Let them and those who stand in authority on the other side of the veil be the judges. *You* just need to fulfill your responsibility to them. Remember your purpose is to act as the Savior did, and I'm sure He didn't approve of some of the things you've done in your life. Right? But He still paid the price for those who would repent. If your ancestor repents and accepts the ordinances, so much the better. You have acted as a "savior" to him/her.

I speak from experience. My Uncle Dan was "a real piece of work" during his life. When he died a year ago, I thought my mother would have his work done for him. However, one thing or another kept her from accomplishing it. So I had my daughter get his baptism and confirmation done along with some other family members that I was "just sure" would accept the Gospel in the Spirit Prison, because they were such good people on the earth. Guess what.

When I was the proxy doing the initiatory work for these people in the Winter Quarters temple, guess whose name brought a strong testimony by the Spirit for each of the four ordinances. My Uncle Dan's! Of all those people, the one that I had least expected to accept the Gospel had been the one to accept it wholeheartedly! So *never* try to judge who should or who shouldn't receive their ordinances. Just do your duty.

HOOKING UP YOUR ANCESTOR

You now have two pedigree "trees" in your PAF file, due to your download from *Ancestral File* in the last chapter. We need to get your ancestor's tree hooked up with yours. The quickest way to accomplish that is by using a PAF tool called **Match/Merge**.

- ❏ Click on the **Tools** menu
- ❏ Click on **Match/Merge**
- ❏ Click **Yes** to make a backup of your file
- ❏ Insert a floppy diskette into the A: drive
- ❏ In the **Save in:** box, click the down arrow
- ❏ Select **3 ½ Floppy (A:)** from the pick list
- ❏ Click the **Backup** button
- ❏ Click **OK** when the backup is complete
- ❏ Click the **Cancel** button on the "Match/Merge Options" screen
- ❏ Click the left **Find** button
- ❏ Click the **Individual List** button
- ❏ Type in the <u>surname</u> of your ancestor (this will take you to that area of the list where you need to look for your ancestor's name listed twice)
- ❏ Select one of the duplicate names for your ancestor, and click **OK**
- ❏ Click the <u>right</u> **Find** button
- ❏ Click the **Individual List** button
- ❏ Type in the <u>surname</u> of your ancestor (this will take you to that area of the list where you need to look for your ancestor's name listed twice)
- ❏ Select the other one of the duplicate names for your ancestor, and click **OK**
- ❏ Compare the information for the two names on the left and right of the screen. If the one on the right looks like it has better (more accurate or complete or <u>reliable</u>), click the **<Switch>** button. This will swap the names around.
- ❏ Now, click the small boxes of any information you want to transfer from the name on the right to the name on the left
- ❏ Click the **Merge** button
- ❏ Click the **Edit** button, if you need to correct any information. Do your corrections and **Save**.
- ❏ Click the **Close** button
- ❏ Click **OK**

Your ancestor's pedigree is now a part of your pedigree! Pretty easy, huh?

DUPLICATES

At this point in this book, you probably have some duplicate names in your PAF file. Any downloads from *Ancestral File* will be sure to create duplicates. It's easy to merge <u>exact</u> duplicates. Here's how:

- ❏ Click on the **Tools** menu
- ❏ Click on **Match/Merge**
- ❏ Click **Yes** to make a backup of your file

- ❏ Insert a floppy diskette into the A: drive
- ❏ In the **Save in:** box, click the down arrow
- ❏ Select **3 ½ Floppy (A:)** from the pick list
- ❏ Click the **Backup** button
- ❏ Click **OK** when the backup is complete
- ❏ Select **Ancestral File Numbers (AFNs)**
- ❏ Check all of the boxes under "Merge", except **Confirm when Merge button pressed**
- ❏ Click **OK**
- ❏ Click **Next Match**
- ❏ Allow PAF to compare and merge all exact duplicates
- ❏ Click **Close**

Whenever you do a Match/Merge, there's always a good chance that you have other duplicate names to clean up before you move on. What happens is this. When you merge a person into another duplicate name, usually the spouse's name goes with it, too. So you end up with two spouses with the same name married to one person! You need to merge those spouse names, also. How do you do that? Here's THE EASY WAY to match and merge duplicate names:

Do **Help Lesson # 6**

After you have done Help Lesson #6, continue to check www.familysearch.org for more downloadable information to build out your pedigree. Eventually, you will exhaust your search options, but this is the easiest way to start out. Always do your match/merge following any download, so that the job doesn't become too big of a burden. If you do it while it's fresh, the task will be much easier to do.

> **Note**: Any download from Ancestral File will almost always contain lots of duplicate names, but don't worry about merging those until after "Trimming the Tree" in the next chapter.

You can imagine that you only want to download names that you really need, so be sure you don't make yourself extra work by getting in too big of a hurry. That happens when you get excited at finding a lot of information. You start downloading *everything*, and then you end up with too much. Think through how you will accomplish the match/merge with the least amount of work.

We want to keep this business as EASY as possible.

Continue entering all *you* know into PAF, *documenting your sources as you go.* Type them in as you work, or you'll forget.

How do you document something downloaded from *Ancestral File*? Well, you don't have to! I told you this was EASY! The Church has already done it for you. You don't believe me? Well, okay. Check this out.

- ❏ Click the **Pedigree** tab and click on the ancestor you downloaded from *Ancestral File*.
- ❏ Click the **"memo pad" icon** on the toolbar. This will bring up your ancestor's "Notes" screen.
- ❏ Click the **Sources** button

What do you see? A nice source note showing that you downloaded that person from *Ancestral File*!!! Could it get any easier?

READ YOUR *MEMBER'S GUIDE*
>>> TODAY<<<
if you haven't done so already.

By now, you should have the beginnings of a family history file with which you can be quite pleased.

KEEP IT UP!!!
You're doing a great job!

CHAPTER SEVEN
"Taking Them to the Temple"

In this chapter, I'll show you what you need to do to create a floppy disk to take to the Temple and do your ancestors' ordinances. By now, you should have quite a few names in your PAF file, all of which need to be checked out to see which ones can be taken to the temple. But before we do that, I want to tell you a personal story. I want to show you, again, how close your dead are to you, once you come to know their stories.

A NEAT STORY

In 1991, when the Church first made PAF and *Ancestral File* available, we were living on the Illinois side of the Mississippi River in the Nauvoo Ward. I downloaded my Call ancestry as a "starter" for my first PAF file, so I had quite a long pedigree on my Dad's side. However, my Mom's pedigree was quite short. (As you will recall, she is the only member of the Church in her family.) She had tried to get information from her family, but hadn't gotten too far, and she had become quite discouraged. What would *you* do, if *your* father only remembered his mother and father's names as "Ma" and "Pa"?

Well, before Sacrament Meeting one Sunday, I was showing off the new, computer-generated printout of my pedigree to a friend at Church, Brother Chuck Allen (who, by the way, manufactured the windows for the new Nauvoo Temple). He noticed that we were cousins, because we had the same Puritan grandmother, way back in Boston. But that's a different story.

Later, as I sat in Sacrament Meeting, gazing upon my mother's short pedigree lines, I noticed that an end-of-line grandmother, Mary Virginia Hill, had been born in Clark County, Missouri, in 1847. Now, in my travels as an employee of the Church, I remembered having passed through Clark County. I remembered the name, but couldn't place where it was, geographically. So when I got home that day, I got out my atlas. Lo, and behold! It was just across the river from where we lived!

My great grandmother was born just across the river in Missouri...one year after the Saints were driven out of Nauvoo. This I had to look into!

It turned out that I'd been driving through my ancestors' homesteads for the past six years! And when I found *that* out, and visited some of their graves, and read of how they died, all of a sudden they became very real—and very close to me. Each time I would drive down those roads where they had lived, they crowded in on me—until I got their temple work done in Chicago. Then, it was as if they had left and gone elsewhere. No longer did they "haunt" those old homesteads, it seemed. They had moved on to higher levels. You may find that happening to you as you learn about your family.

In regards to our dead, Joseph Smith taught that "*they are not far from us*, and know and understand our thoughts, feelings, and emotions, and are often pained therewith."[18] Parley P. Pratt wrote that the spirit world "is here on the very planet where we were born."[19] And, lastly, Brigham Young preached in 1856,

> "When you lay down this tabernacle, where are you going? Into the spiritual world...Where is the spirit world? It is right here. Do the good and evil spirits go together? Yes, they do. Do they both inhabit one kingdom? Yes, they do. Do they go to the sun? No. Do they go beyond the boundaries of the organized earth? No, they do not. They are brought forth upon this earth, for the express purpose of inhabiting it to all eternity. Where else are you going? No where else, only as you may be permitted."[20]

HELP LESSON #7

Now is the time for you to do Help Lesson #7. It will give you the insight you will need for the remainder of this chapter.

PREPARING ANCESTORS FOR THE TEMPLE

Now that you've done Help Lesson #7, let's get started preparing your ancestors' names for the temple. It's easy (as always). But, first, you may want to do some "house cleaning". You may want to "trim your tree" (your pedigree tree, that is). In your collaborations with other family members, or in downloading from the Internet, you probably have picked up some extraneous names—names of people whom you aren't responsible to do the ordinances for. You know who I mean—second and third cousins and the

---

[18] *Teachings of the Prophet Joseph Smith*, pg. 326, emphasis added

[19] Pratt, Parley P., *Key to the Science of Theology*, 10th ed., Salt Lake City, Deseret Book, 1948, pg. 130.

[20] *Journal of Discourses* 3:369

like. You may want to eliminate those names from your PAF file. You do that by creating a new, "lean and mean" file.

"TRIMMING THE TREE"

I call this "trimming your tree" (your pedigree tree). If you want to clean up your PAF file, so that it only contains those ancestors for whom you are responsible, you need to follow my instructions in Supplement E. In effect, you will be creating a whole new file, minus the names you really don't need. That will be the file that you will pull names from to take to the temple. However, it's a good idea to keep the old file around, just in case you need it in future research.

Now is the best time to go through your new PAF file and merge any duplicate names. It will take you some time, so do it only when you have a few hours to concentrate on just that. Believe me, Match/Merge is not a place where you want to make mistakes! They are hard to undo. You will have to make decisions based upon good sense and what the Spirit guides you to do, so please pray for guidance before doing this work. It will make all the difference in the world, and it will make it so much easier.

CONTRIBUTING TO THE ANCESTRAL and PEDIGREE RESOURCE FILES

In Help Lesson #5, you should have learned how to contribute *your* information to the Church's two master databases,

Ancestral File
and
Pedigree Resource File

I shouldn't have to repeat that exercise here. However, I *do* want to strongly recommend that you be a *giver*, as well as a receiver. By now, I hope you have been filled with gratitude for the Church's efforts in providing you with free information. There are many other databases that you can also search for your ancestors. We'll discuss those in the next chapter.

Please understand that these two main databases are the real "workhorses" for FamilySearch. Without them, we would still be in the dark ages of genealogy. They are the gifts of a loving God, and you can give, also, by sharing *your* information with the Church. Then other folks can use that information to do further research, which may eventually open the door for *you* to go even farther back on *your* lines. The whole key to genealogical research, nowadays, is sharing.

CORRECTING ANY PROBLEMS WITH YOUR DATA

After trimming your tree, before you go any further, you need to check the data in your database to see if there are any weaknesses that need repair. You can do that by running **Check/Repair**, as shown in Supplement G.

PREPARING FOR TEMPLEREADY

With your database all cleaned up, you are now ready to begin the process of qualifying names to take to the temple. You do this on your own computer, and then you take the disk that is created (what I call a "Temple Prep disk") down to your local Family History Center to run it on the TempleReady software. The outcome of that process will, hopefully, result in a disk with names qualified to take to the temple. The whole process is diagrammed in Fig. 4.

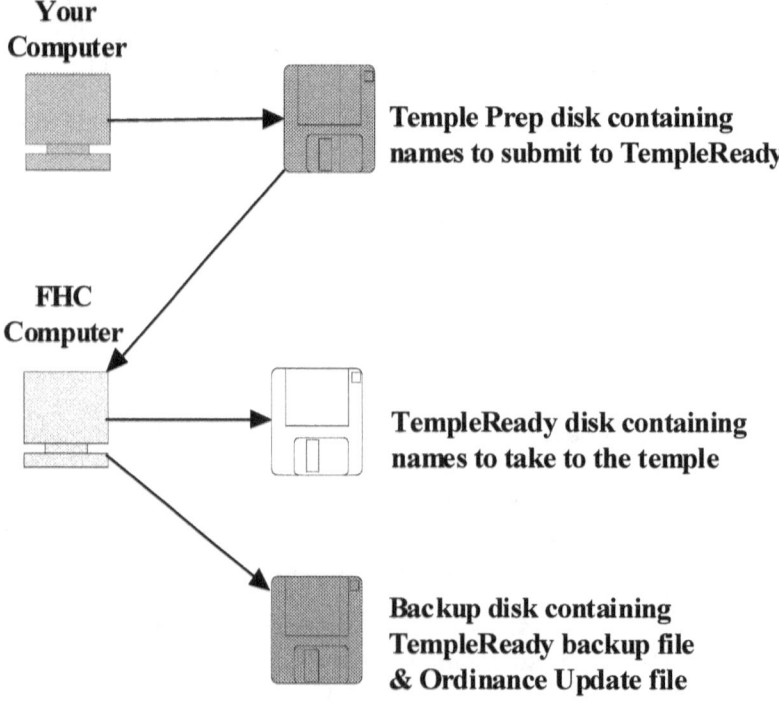

Your Computer

Temple Prep disk containing names to submit to TempleReady

FHC Computer

TempleReady disk containing names to take to the temple

Backup disk containing TempleReady backup file & Ordinance Update file

Fig. 4

PROCESS OF QUALIFYING NAMES FOR THE TEMPLE

The first step in this process is to follow the instructions given in Supplement F. This will create a Temple Prep disk. After you have that disk in hand, take it down to your local FHC and get help on running their TempleReady software to create the disk you will take to the temple. Also at the FHC, you will create a disk with a backup file of the one you will be taking to the temple. So you will come away from the FHC with at least three disks:

- Your Temple Prep disk
- The disk you will take to the temple (with a special label available from the FHC staff)
- A backup disk (with a special label)

TEMPLEREADY ORDINANCE UPDATES

If your FHC has the latest version of TempleReady for Windows, and your version of PAF is 5.2.18 or newer, one of the closing dialog boxes in TempleReady will ask you if you want to create a temple ordinances update file for your PAF. Always check that box. It will save you a lot of work. TempleReady will ask you for another floppy disk for the Ordinance Update file. So you may leave the FHC with *four* disks. If you have version 5.2.18 (or newer) of PAF, you can take that Ordinance Update disk home to your computer and update your PAF file with the already-completed ordinance information found by TempleReady. You can do that as shown in Supplement H.

Slick, eh? Nowadays, TempleReady and PAF "talk" to each other!

BACKUPS

As always, keep your backup disks separate from the disks you will take to the temple. Believe it or not, sometimes a temple disk *does* get damaged, and you have to use your backup disk. So keep them safe and sound, so you won't be tearing your hair out later!

Are you sweating yet?
I didn't think so.
This stuff is EASY!

Remember the goal?
Well, you're almost there.

WHAT IF SOMETHING GOES WRONG?

There are times when you take your disk down to the FHC and something goes wrong with the processing of your file. If your file is just unacceptable to TempleReady, or if one name botches up the whole process, you have the possibility of a major problem on your hands.

When you created your Temple Prep disk, using Supplement F, you told PAF to put the word "Submitted" in each ordinance field that qualified. The reason we did that was so PAF wouldn't pick those people up again for ordinance work. However, if you have to go back and re-create a Temple Prep disk, PAF won't pick the same people up again. So you have to have PAF erase the "Submitted" designation that it inserted in those fields.

The way that is done is as shown in Supplement I. Then you can go in and create another Temple Prep disk for the same names as before.

CHAPTER EIGHT
"Reports and Research"

"Long genealogies and endless pedigrees are a sorry offset for short memories and shallow brains." [21]

Remember Fig. 2 in Chapter Three? If you go back and look at that figure, you will see that, at the beginning of Chapter Seven, you were at Step 4. By the time you finished Chapter Seven, you had leaped to Step 8! That's a pretty big jump, and it's all due to the new PAF program. We were able to replace Steps 5 and 6, with just one step—Prepare a Temple Preparation Disk.

Now that you have your names ready for the temple, all you have to do is get the ordinances done by following the remaining steps in Fig. 2. See if the YM/YW group in your ward/branch will help you do the baptisms and confirmations. Better yet, have your children or grandchildren do them. Involve your own family as much as possible. Baptisms, confirmations, initiatories, and sealings go pretty quickly. It's the endowments that take the time. Enlist as much help as possible from family and ward members. Remember, not only are they helping you out, they are blessing the lives of your ancestors *and* blessing their own lives, at the same time.

Make a diligent effort to keep track of who has been assigned each name, or you will never be quite sure whether the work has been done. However, the Church has just begun a new service on their www.familysearch.org website. If you have your membership number and your confirmation date, you can sign in and get up-to-date ordinance information on any of your ancestors. So, if you lose track of who is in possession of a name with "Submitted" in an ordinance field, at least you can check to see if the ordinance was completed. Instructions for doing that are in Supplement J.

When you take your TempleReady disk to the temple office, the staff will use it to print out the pink, blue and yellow name slips for you. Additionally, they will give you a "tracking sheet", showing all of the names and check boxes for their ordinances. This tracking sheet is what you should use to keep an account of who has what name slips. Don't ever send your TempleReady disk to someone else. You need to keep all disks and tracking sheets in your possession. Only send name slips through the mail.

---

[21] Edmund Fairfield, President of Hillsdale College, July 4,1853

There are just too many ways for "Old Scratch" to upset what you're trying to do for your ancestors. By the way, dedicate your home, so he won't know where you keep your disks. If you don't know what I'm talking about, talk to your bishop/branch president/home teacher.

NIFTY REPORTS

PAF has some really useful reports that you can print out. For example, you can have it check your database for Possible Problems, or you can print out all of your End-of-line Individuals. Now is the time to do Help Lesson #4. Then you will be an expert on "reports."

INTERNET RESEARCH RESOURCES

One of the best, new resources that the Lord has made available to genealogy researchers is the Internet. Lots of information is now available online that used to require trips to local county courthouses and historical societies. Nowadays, there are many historical societies that are working hard to save us time and money.

Genealogical websites are numerous, and they are increasing by the day. Genealogy is said to be the second most popular hobby in the country. The number of websites seems to bear that out. There are websites that exist strictly to link you to other genealogy websites. There are those that are trying desperately to put census information online. Some even give you actual images of the census taker pages! The great majority of the websites require some money from you, however there are many that are totally free. My favorites are the Church's www.familysearch.org, along with www.usgenweb.org and www.gendex.org. RootsWeb has a portion of their site that is free at www.rootsweb.com, and if you click on their *World Connect* link, you may find some really good stuff. I've included a list of useful websites in Supplement K. Check them out. You may find your lost relative lurking there.

"CATCHERS' MITTS"

Remember how, in Chapter Two, I promised to discuss the differences between LDS and non-LDS genealogy methods? I said that non-LDS family trees tend to be backwards from ours—and that's good—for us. Whereas our family trees are "ancestor-driven", theirs tend to be "descendant-driven". See Figure 6.

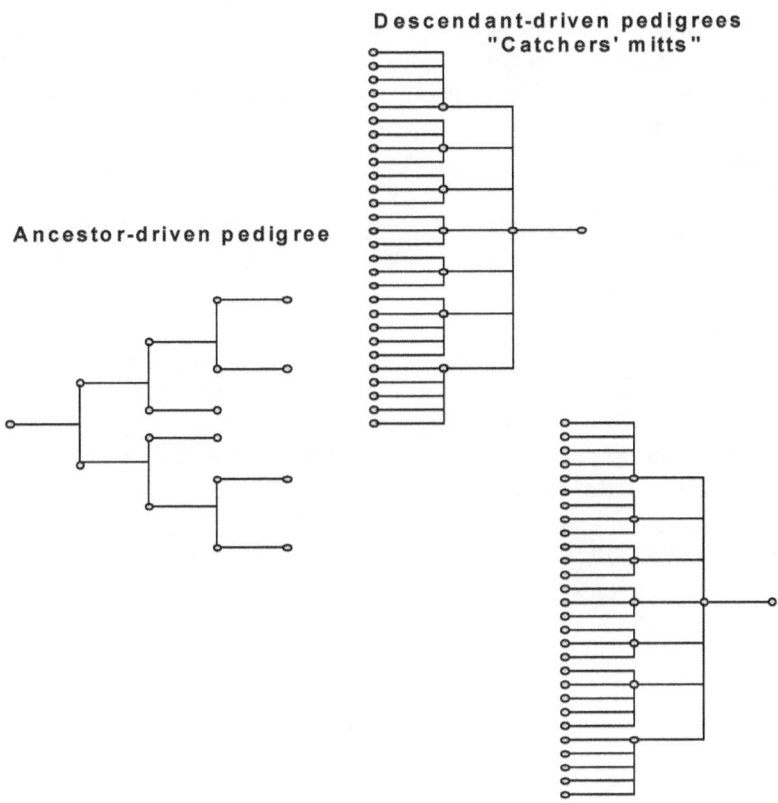

Fig. 6
LDS vs. non-LDS Family Trees

In other words, non-LDS folks seem to have a great desire to find all of the *descendants* of one of their ancestors, where we are only really interested in *our ancestors*.

Our pedigrees are shaped like the one on the left, while theirs are shaped like the ones on the right. This creates a lot of what I like to call "catchers' mitts" out there. And they make it a lot EASIER for us to get our ancestors' ordinances done. If one of *our* end-of-line individuals happens to be one of *their* descendants, we can download their GED, make the connection, and "we're in business." Isn't it amazing how the Spirit of Elijah works?

RESEARCH OUTLINES

The Church has made state research outlines available for years at just the cost of the printing, but now such guides are usually available to you online.

They are wonderful, state-by-state guides that tell you where each specific state of the union has its records archived, where the historical societies are located, and where to look for specific types of records. They usually even give you a handy chart telling you where to look *first*. You can locate them online at:

http://www.familysearch.org/Eng/Search/RG/frameset_rhelps.asp

RESEARCH GUIDANCE

Another wonderful tool that the Church has made available to us online is the *Research Guidance* portion of *FamilySearch.org*. From http://www.familysearch.org/Eng/Search/RG/frameset_rg.asp you can click on any state or country and be counseled on research strategy for any type of event you want to find.

THE RESEARCH "PUZZLE"

Doing family history research is very much like working a picture puzzle — usually no one piece tells the whole story. It normally takes several pieces of information that "fit together" before you can be sure that all of the "pieces" belong to the same ancestor. That's what you are trying to do in your research — **prove to yourself that a person actually lived at a certain time in history, in a certain place, and in a certain family**.

Our family once lived in the "Gold Country" of California, not too far from where gold was first discovered at Sutter's Mill. The '49ers who *first* came to the gold fields of California were basically "high grading" the *easy* yellow stuff. It was easily picked up, lying around in streambeds. But later comers had to work hard for their gold.

That's pretty much what you've been doing up to this point in your research (gold digging). You've been "high grading" information from the EASY sites like Ancestral File and Pedigree Resource File. Once you have done all the easy stuff, if you haven't completed all of your six-generation family groups, you will need to work on your "end-of-line individuals." That will mean printing out a list of those names, as taught to you in Help Lesson #4.

Once you have your list, look at your pedigree chart. Usually the end-of-line individual who is farthest left on the chart will be the easiest to find, so start with that person. The women in the family groups are always the hardest to find, due to the loss of their maiden name with marriage, *but they must be found*. If you have a family group with only boys as children, *there is a good chance that you have some girls to find,* also. Families were usually big in the "old days", so the odds of having all boys would be "a long shot".

CENSUSES

Even though they are not considered to be primary evidence by professional genealogists, I feel that anyone who neglects looking in all of the pertinent censuses at the beginning of their search for an ancestor is going to really make it hard on himself. And this book is all about how to make this stuff EASY. So begin your search by going to the correct censuses, both federal and state. The later the census, the better. The 1880 U.S. Census is my favorite, and it can be searched for free on www.familysearch.org, as can the 1881 Canadian and English censuses. The U.S. census gives not only the birthplace of the person enumerated there, but also *the birthplaces of their parents.* So, if the person you're looking for could be in the United States in 1880, that's the place to start your search.

MY CHART

As you work your way backward in time, using censuses and marriage records, you will pick up lots of names. Many will look like the same person, but you should *never assume* that they are. Only when the "pieces of the puzzle" prove to you that they are the same person, should you ever do so. Therefore, you must keep them separate until that time.

I'm a *visual* person. The more you can show me something in pictures or diagrams, the easier it is for me to understand it. So I had to develop a "picture" of the names I was researching. I decided that a "bar chart" would work best, and I developed a way to keep track of names that showed up on censuses every ten years. I use easily purchased graph paper, and Fig. 5 is an example of one of my research charts. The top name is always the end-of-line individual I'm concerned with. The dates of the censuses go along the top of the sheet in ten-year increments. If I definitely know of the date of an event, it gets a vertical bar and a brief note as to the location of the event. To me, a census is an event, and it gets a note of where it was taken. Marriages are easily added, as you can see.

From a census record in 1850, I will write the person's name and, at the "1850" vertical line on the graph paper, I draw a short horizontal line. Then I extend that line to the right however many years that person is old, and make a vertical mark in the year that they were supposed to have been born. Remember, this is not an *accurate* chart. It's just to get a visual picture of how families *may* fit together.

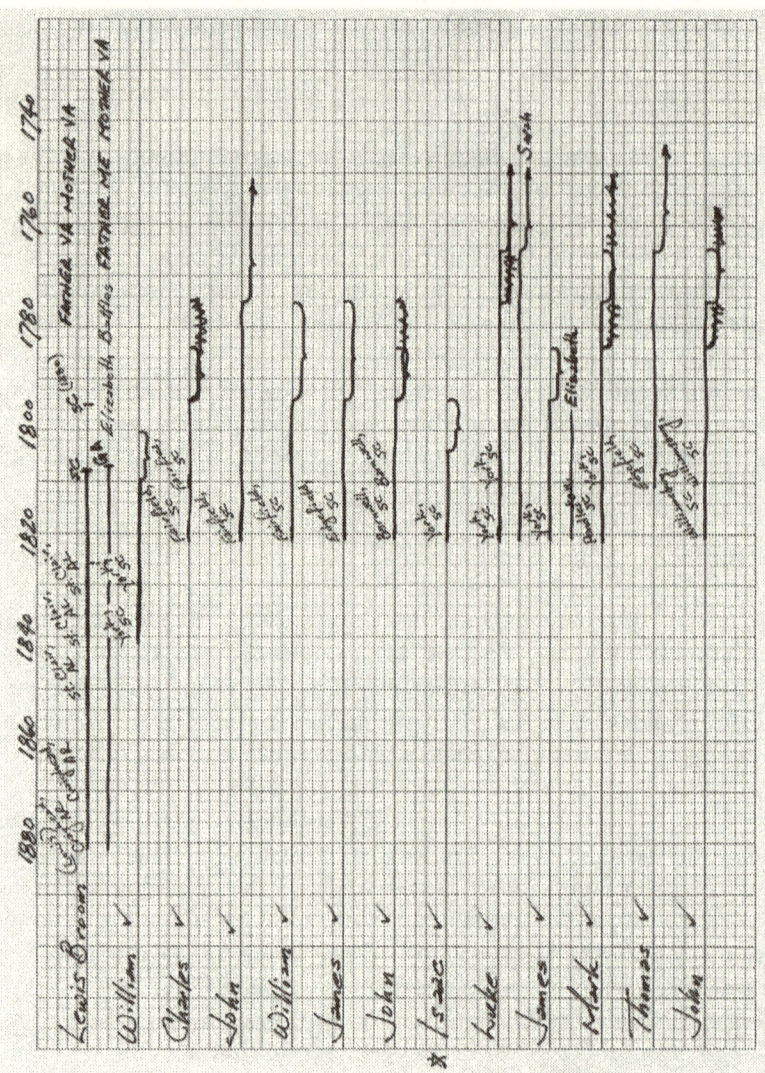

Fig. 5
My Chart

If I am looking at a census that only gives the head of household, then people in the household are usually shown in age *groups*. So I put the head of family name on the line, drawing my horizontal line from the census year to the right until I reach the lowest age of the age group that he is in, then drawing a horizontal *bracket* for the entire age group. If a father stays in one place long enough for several censuses to be taken with him as the head of a family, those brackets can be superimposed on each other — thus narrowing the "target" years of his birth. Also, new family heads with the same surname may start to show up nearby on the census taker's sheet. If they are young enough to be sons of the original head of family, they probably are. Sometimes married daughters tend to get "hidden" next door, also, under their married names.

Always keep in mind that "family" was extremely important on the frontiers of early America. Families migrated *together*, not singly. Often they had close friends also with them — other families that sons and daughters often married into. All of these bits of information must be kept in the researcher's mind or on paper. Eventually, they are all brought together, though, *in the mind*, like pieces of a puzzle. If you like doing puzzles or reading mysteries (trying to figure out "whodunit" before the author tells you), then you may have the verve and stamina for this kind of *fun*. But, if you don't, let someone else in the family do the research, and you do the temple work. Of course, if you're the only member of the Church in your family, there may be a shortage of those kinds of people you can turn to. However, the Spirit of Elijah does infect even non-members, so you may be in luck.

CONCLUSION

I told you in Chapter Four that I would give you just a *smattering* on how to do research. I feel I have done that. From here on in, you're on your own. If you do your own research, there is one book that you should buy, because it is invaluable in understanding the censuses you will look at. That book is the *Map Guide to the U.S. Federal Censuses, 1790-1920*, by Thorndale and Dollarhide. It's impossible to make sense out of county boundaries without it. A new website is now available that is also quite helpful in this vein,

www.genealogyinc.com/maps/uscm.htm

And another good book for understanding *how* to research is *The Source*, by Szucs and Luebking.

My purpose in writing this book was to (1) teach you the doctrine of work for the dead, (2) teach you how to create names for the temple, and (3) give you a smattering on how to do the EASY research. I feel that I've accomplished that, so it's time to bring this book to an end.

I hope you've enjoyed reading it as much as I've enjoyed writing it. These are the things I've been teaching in my Family History Sunday School classes for the past ten years. You should now be at the same level as any of my students as they leave my classes.

May the Lord bless you in this great work upon which you have embarked, and may your ancestors open the doors for further work for them. Sometimes it just takes patience and a few years of waiting before those end-of-line individuals are able to bring about the information necessary.

I remember the story of a good sister who was completely frustrated with one of her family lines. She just couldn't make any more headway on it. Disgusted, she threw the file into a bottom drawer, kicked it shut, and said, "There! If you want to get out, you're going to have to help!"

And they did!!!

<u>SUPPLEMENTS</u>

Chuck Call

SUPPLEMENT A

"HOW DO I START...?" BROCHURE

- ❑ Go through all of the steps in the *How Do I Start My Family History?* brochure **FIRST**, filling out as much of the pedigree chart as possible.

DOWNLOAD AND INSTALL PAF 5.2

- ❑ Go to http://www.familysearch.org/
- ❑ Click **Order/Download Products**
- ❑ Click **Software Downloads - Free**
- ❑ Click the **Download** button for **Personal Ancestral File 5.2.18.0 - Multi-Language (9.7 MB)**
- ❑ Click the **Continue** button at the bottom of the screen
- ❑ Fill in your Registration information, and click **Send**
- ❑ Click the **FTP Download** button by the **English (9.7 MB)** version
- ❑ *RealPlayer* has been known to cause problems when downloading PAF, so if you have it on your computer, disable it while downloading.
- ❑ If you have problems with the download, try the **HTTP Download**
- ❑ Check "Save this program to disk" and click **OK**
- ❑ Save download to **Desktop** in the "Save in:" box, and click **Save**
- ❑ *Be patient while the download takes place.* It may take some time!
- ❑ When download is complete, click **Close**
- ❑ Now install the PAF program by opening your desktop and clicking the icon named

- ❑ **PAF5EnglishSetup.exe**

- ❑ Select **English** and click **OK**
- ❑ Follow the remaining installation instructions, **having an icon placed on your desktop**

SUPPLEMENT B

HELP LESSON #1

- ❑ First, you must be connected to the Internet
- ❑ Click the **PAF 5.2** icon on your desktop
- ❑ Click the **Cancel** button in the opening dialog box
- ❑ Choose the **Help** menu
- ❑ Click **Lessons**
- ❑ Read the opening instructions and download the Macromedia Flash Player, if necessary
- ❑ Click **1 Getting Started: Typing in Your Family** under "Lessons" in the left sidebar
- ❑ Click **1** in the numbered grid that opens. This will take you to "Create a PAF file"
- ❑ Follow the instructions
    - o Note that what you are looking at is a "Help" window
    - o It is NOT the window you will be working in. It is only the instructions for working in the PAF window.
    - o In other words, you have two windows open at this time...the PAF window and the HELP window opened from the PAF window. The HELP window gives the lesson instructions for your work in the PAF window.
- ❑ Work your way, step-by-step, through the numbered grid, until you complete the lesson
- ❑ Create two backup disks, not just one, and store them in separate locations...for safety's sake.

SUPPLEMENT C

HELP LESSON #2 (also used for Lessons 3 thru 7)

- ❏ First, you must be connected to the Internet
- ❏ Click the **PAF 5.2** icon on your desktop
- ❏ Click the **Help** menu
- ❏ Click **Lessons**
- ❏ Click **2 Getting Started: Making Changes** under "Lessons" in the left sidebar
- ❏ Click **1** in the numbered grid that opens. This will take you to "Family View"
- ❏ Follow the instructions
  - o Note that what you are looking at is a "Help" window
  - o It is NOT the window you will be working in. It is only the instructions for working in the PAF window.
  - o In other words, you have two windows open at this time...the PAF window and the HELP window opened from the PAF window. The HELP window gives the lesson instructions for your work in the PAF window.
- ❏ Work your way, step-by-step, through the numbered grid, **clicking on all of the neat things you will** see, until you complete the lesson
- ❏ Create two backup disks, not just one, and store them in separate locations...for safety's sake.

SUPPLEMENT D

DOWNLOAD FROM THE INTERNET ANCESTRAL FILE

- ❑ Go to http://www.familysearch.org/
- ❑ Click **Search for Ancestors**
- ❑ Type in your ancestor's first and last names
- ❑ Hit the **Enter** key on your keyboard
- ❑ Look for your ancestor among the lists of names that are returned to the screen
  - o Look, first, in the "Ancestral File" list
  - o Then, look in the "Pedigree Resource File" list
- ❑ If you don't find your ancestor, try another farther back on your pedigree, or try one from another family line on your pedigree.
- ❑ Once you find an ancestor that appears to match the one on your pedigree, click **Pedigree** to the right of his/her name. This should return their pedigree to the screen. If you're lucky, you may hit a gold mine!
- ❑ If their pedigree seems to be what you want, it's time to download that pedigree.
  - o NOTE: What you see is what you get (WYSIWYG). The Internet Ancestral File is different from the CD version, which you will find at your local Family History Center. The Internet version is WYSIWYG (said "whizzy-wig"). In other words, you can only download the "chunk" of a pedigree that you actually **see** on your computer screen! If the pedigree you're looking at has "right arrows" at the ends of some of its family lines, then you can click on those arrows, one-at-a-time, to bring to your screen additional "chunks". It's slower, but it's a lot more up-to-date information. The CDs at your FHC could be two years old. The Internet site is as fresh as it gets!
- ❑ From your ancestor's pedigree screen, click on **Download GEDCOM**
- ❑ Click **Save this file to disk**, then click **OK**
- ❑ In the **Save in:** box, click the down arrow and select **Desktop** from the pick list
- ❑ Click **Save**
- ❑ An icon with your ancestor's surname under it should have appeared on your computer desktop
- ❑ You now have completed the download of your ancestor's pedigree

SUPPLEMENT E

HOW TO TRIM YOUR PEDIGREE TREE

- ❑ Click on your own name
- ❑ Click on the **File** menu
- ❑ Click **Export**
- ❑ Click **Partial**
- ❑ Click **Select**
- ❑ In the "Relationship Filter" box, click the down arrow, and select **All Ancestral Related** from the pick list (*NOTE:* This will pick up all family groups on your pedigree chart, plus the children's spouses, but will not pick up cousins)
- ❑ Click the **Select** button
    - o If a question is returned to your screen asking about "all parents," click **Yes**
- ❑ Look at the list of names in the box. All selected names will have >> next to them.

To drop names from your list:
- ❑ Change the "Relationship Filter" to **Individual**
- ❑ Click on the name you want to drop
- ❑ Click **Deselect**

To add names to your list:
- ❑ Change the "Relationship Filter" to **Individual**
- ❑ Click on the name you want to add
- ❑ Click **Select**

- ❑ When your list looks the way you want it, click **OK**
- ❑ Make sure all the boxes on the right side of the screen are checked
- ❑ Click **Export**
- ❑ Give your file a name
- ❑ Click **Export**
- ❑ Click **OK**
- ❑ Click **OK**

SUPPLEMENT F

HOW TO MAKE A TEMPLE PREPARATION DISK (PAF 5.2)

- ❏ Open your PAF file
- ❏ Click on your own name
- ❏ Click the **File** menu
- ❏ Select **Export for TempleReady for Windows**
- ❏ Click the **Continue** button
- ❏ In the *Relationship Filter* box, click the down arrow
- ❏ Choose **Ancestors** from the pick list
- ❏ Click the **Define** button in the *Field Filter* area
- ❏ Scroll down to the bottom of the list, and choose **Qualified for Baptism/Endowment**
- ❏ Click **>**, then check **Is** and click **OK**
- ❏ Click the **AND** button
- ❏ Choose **Birth Date** from list, then **>**
- ❏ Click the down arrow in the *Options* box
- ❏ Choose **Is greater than** from pick list and enter **1600** in the *Date* box
- ❏ Click the **AND** button
- ❏ Choose **Birth Date** from list, then **>**
- ❏ Click the down arrow in the *Options* box
- ❏ Choose **Is less than** from pick list and enter **1890** in the *Date* box
- ❏ Click **OK,** then click **OK** again
- ❏ Check the **Show results only** box
- ❏ Click **Replace** in *Relationship Filter* area, then **Yes** (if a warning is displayed)

To drop names from your list:
- ❏ Change *Relationship Filter* to **Individual**
- ❏ Click the name you want to drop
- ❏ Click **Deselect**

To add names to your list:
- ❏ Uncheck **Show results only**
- ❏ Change *Relationship Filter* to **Individual**
- ❏ Click the name you want to add
- ❏ Click **Select**

❑ When your list looks the way you want it, click the **OK** button

❑ Check the options you want, making sure to check the box that says "Put the word 'Submitted' in ordinance fields..." You will also want a submission report.

❑ Click the **OK** button

❑ Insert a floppy disk into Drive A, then, on the dialog window, select the **3 ½ Floppy (A:)** drive and give your file a name.

❑ Click the **Export** button

❑ Click **OK**

❑ Read the instructions, and click **OK**

❑ Your submission report will pop up on your screen, you should print this out.

❑ Click the **File** menu

❑ Select **Print**, then click the **OK** button

SUPPLEMENT G

HOW TO RUN CHECK/REPAIR

- ❑ Open your PAF file
- ❑ Click on the **File** menu
- ❑ Select **Check/Repair**
- ❑ Click the **Check** button
- ❑ Click the **Save** button
- ❑ PAF will then check all of your data
    - o If errors are detected, PAF will give you an on-screen report to read. Using the RIN and MRIN, check your data to determine where the problem(s) may be and make correction(s).

    - o If you can't figure out how to correct a problem, PAF will try to correct it, if you run **Check/Repair** again, and click the **Check/Repair** button, instead of the **Check** button.
        - ▪ CAUTION: This method for correcting data could result in the loss of some of your information, but sometimes it's the only way to fix it.

- ❑ If no problems were found, click **OK**

SUPPLEMENT H

HOW TO ADD TEMPLEREADY ORDINANCE UPDATE

- ❑ Open your PAF file
- ❑ Insert the Ordinance Update disk in drive A:
- ❑ Click the **File** menu
- ❑ Select **Add TempleReady Update**
- ❑ Click the down arrow in the *Look in:* box
- ❑ Select **3 ½ Floppy (A:)**
- ❑ Select the file you created, and click **Add**
- ❑ Wait while the ordinance dates are added to your PAF file.
- ❑ Certain "Sub [date]" fields will be replaced by the actual ordinance information from TempleReady
- ❑ When the process is finished, a *Results of TempleReady Update* report will appear. (This file has the same name as the original PAF database with a .lst extension.) This report lists all of the RINs and MRINs that were updated, and what information was updated. You can print this report by:

  - o Clicking on **File**, and
  - o Selecting **Print**

  or just close the window.

SUPPLEMENT I

REMOVING "SUBMITTED" FROM ORDINANCE FIELDS

- ❏ Open your PAF file
- ❏ Click on the **Tools** menu
- ❏ Select **Global Search and Replace** from the menu
- ❏ When the dialog box opens, click the down arrow, and select **Dates** from the pick list
- ❏ In the *Search for:* box, type **Sub [date of submission]** (for example, "Sub 24 Feb 2003")
- ❏ Leave the *Replace with:* box blank, and be sure the *Case sensitive search* box is checked
- ❏ Click the **Replace** button
- ❏ All dates in your database matching that submission date will be erased.

SUPPLEMENT J

TO CHECK TEMPLE ORDINANCES ONLINE

- ❏ Obtain your Church membership number and confirmation date from the Ward Clerk.
- ❏ Go to www.familysearch.org on the Internet

For first time users:
- ❏ Click "My Info"
- ❏ Click on "new user" link at end of first paragraph
- ❏ Enter your personal information
- ❏ Click "Register" button

For users who have already given their personal information:
- ❏ Click "Sign On"
- ❏ Enter User Name and Password
- ❏ Click "Sign On" button
- ❏ Click "Search for Ancestors"
- ❏ Click "International Genealogical Index"
- ❏ Enter the person's first and last names (a middle name may also be entered after the first given name)
- ❏ Select a "Region" of the world where the person lived
- ❏ Click "Search" button
- ❏ Click on the person's name that fits your search parameters
- ❏ You should see the ordinance work that has been done for that person.

> **TIP:** Sometimes people have had ordinance work done under different name spellings, so you need to check more than one name. The same name might be listed more than once, with different ordinances done for each listing.

SUPPLEMENT K

USEFUL (sometimes) GENEALOGICAL WEBSITES

www.familysearch.org (LDS Family History site)

www.usgenweb.com (USGenWeb site)

http://www.us-census.org/inventory/inventory.htm (USGenWeb's census index by state and county)

http://www.gendex.com:8080/display?page=surnames& (GenDex surname search site)

http://familyhistory.byu.edu/ (BYU's Center for Family History & Genealogy)

www.rootsweb.com (RootsWeb's site)

http://www.rootsweb.com/~rwguide/lesson5.htm (RootsWeb's Guide to Marriage Records)

http://ssdi.genealogy.rootsweb.com/ (RootsWeb's Social Security Death Index)

http://resources.rootsweb.com/cgi-bin/soundexconverter (RootsWeb's Soundex Converter)

http://www.cyndislist.com/ (Cyndi's List — categorized & cross-referenced genealogical web links)

http://www.sos.state.il.us/cgi-bin/archives/marriage.s (Illinois Statewide Marriage Index, 1763 — 1900)

http://www.rootsweb.com/~vagenweb/ (Virginia's GenWeb page)

http://familytreemaker.genealogy.com/?Welcome=1047583242 (FamilyTreeMaker online)

http://genealogy.com/index_r.html (Genealogy.com website)

http://genforum.genealogy.com/ (Genealogy chatrooms and bulletin boards connection)

http://www.bcgcertification.org/associates/index.php (this site allows you to search for a certified genealogist in any area of world)

http://www.archives.gov/facilities/ (a clickable map of the National Archives and the records stored at each location)

http://dar.library.net/ (Daughters of the American Revolution Library)

http://quickfacts.census.gov/cgi-bin/lookup?state=29000 (find out what county a town is in)

http://www.feefhs.org/ (Federation of East European Family History Societies)

http://www.genuki.org.uk/ (United Kingdom & Ireland genealogy site)

http://www.worldgenweb.org/ (The worldwide GenWeb format)

http://www.onegreatfamily.com/ (an online genealogical service)

http://www.geocities.com/heartland/acres/8310/
(GenSearcher...offers...one-stop, on-line research)

www.glorecords.blm.gov (database of federal land conveyance records of public states. Has image access for federal land title records for Eastern public Land States. This site does not include the original colonies and territories.)

www.phmc.state.pa.us (Pennsylvania State Archives. Includes digital images of Revolutionary War military abstract card file, World War I service medal applications cards, Spanish American War Veteran's card file of US volunteers)

www.ellisislandrecords.org (database of 22 million immigrant names from 1892 to 1924)

www.lva.lib.va.us (Library of Virginia includes a many online images, bible records, land patents and grants, etc. Use the site map)

http://www.cyberdriveillinois.com/departments/archives/archives.html (Takes you directly to the State of Illinois Archives website) www.sos.state.il.us/departments/archives/archives.html (another link to the same information)

http://worldconnect.rootsweb.com/ (Surname search)

http://www.iowahistory.org (Iowa State Historical Society Libraries with online catalog)

http://www.censusfinder.com (a directory to FREE census records online)

www.genealogyinc.com/maps/uscm.htm (US & State County Census Maps: Good site offering many county-formation maps, as well as an interesting animated presentation of how the borders changed during each decade)

www.documentsonline.pro.gov.uk (English documents: Good site for searching for English ancestors. Thorough, easy-to-use search engine, and PDF copies of documents can be sent to you by email)

INDEX

INDEX

NOTES

ABOUT THE AUTHOR

Chuck Call is an alumnus of Brigham Young University. Since the advent of the personal computer in genealogy, he has been deeply involved in his own family history work, and has taught Family History Sunday School classes in various wards in the Midwest for the past ten years. With a background of 12 years in family history research and 18 years in project management, his book is well-organized, informative and, best of all— **exciting!** He believes that genealogical work *should* be exciting and not just the arena of "old folks." And he believes the Savior meant what He said— His yoke is meant to be

EASY